Hot Dogs
and
Healing

The Uncomplicated Truth

When we remove our focus from the self
And acknowledge universal truth
We see every moment as an
Opportunity for miracles

Richard G. Visone, D.C.

Dedicated to my perfect wife, Tarina and our beautiful children, Maresca, Camille, Victoria and Giovanni

ISBN   978-0-6151-7923-0

Contents

Foreword

From an early age, I have been absorbed in science, metaphysics, religion and the teachings of Eastern philosophy. I studied the origins and applications of various martial arts, comparing linear techniques to those based upon centripetal motion, and realized the merits of pure form and the unlimited potential of combined forms. Each approach was especially effective in certain ways and the masters consider even the slightest variations of the techniques to be unacceptable. But, those who know more than one technique and when each is most appropriate to apply can approach more varied circumstances with maximum effectiveness.

My experience was similar with every topic I explored. The rules were firm and unyielding and I found huge gaps between what seemed to be related areas. Every theory would only explain so much, giving it distinct limitations and raising questions as to how one may relate to another. It appears that the answer to these questions may be found in the void existing between the boundaries of each premise.

In Newtonian physics, the motion and interaction of matter is examined. That which can be seen and/or quantified is all that exists, and in an attempt to expand the boundaries of this study of such a complex system, the physical engineers have resorted to creating 'invisible' particles. Meta-physics addresses energy and forces which are generally unseen. Through quantum physics we are beginning to enter and explore the void between these two perspectives. Fundamentally, it is the study of the interaction of electromagnetic energy and matter, waves and particles, the path through which universal/innate intelligence may express itself.

I have read many books on physics and metaphysics and most were either written from an analytical (left brain) perspective that explained the rules and outlined the extent of acceptable variation, or they were so abstract (right brain) that the concepts lacked cohesion. But, even the most ancient philosophies teach us that there are no absolutes or limits to variation and that every occurrence, everything that exists, is somehow intimately linked.

It is time we bring philosophy and science together to attain a greater understanding of the 'Big Picture'. I hope that this book will assist you in understanding the underlying causes for many social, psychological and physical problems by expanding your awareness of who you are and what is going on around you. It may also help health care providers to effectively integrate holistic principles with science knowledge to become pioneers of quantum healing and powerful facilitators for maximizing human potential and the physical experience.

Introduction

You may find many of the concepts to be presented in this book difficult to embrace. The reality we will explore may differ from your perceptions of truth simply because it contradicts what you have been taught to believe. Moreover, the three-dimensional conscious reality we create is a polarized world, a world of black and white in a universe of infinite color and possibilities.

The purpose of this writing is to simplify and unite scientific and etheric principles into an easy to understand and practical system which will clarify your perception of truth and reality and, consequently, improve the quality of every aspect of your life. I could clearly write an entire book on each of the several topics and sub-topics addressed in the following pages. However, my goal is to avoid going into too much technical detail and to present basic concepts that will provide a foundation for expanded awareness and enlightened thinking. Wisdom is an acknowledgement of reality…the recognition of truth.

Keep in mind that this form of communication is, quite literally, black and white. The absolute truth is…nothing is absolute! Due to the inherent polarity of the universe, the duality created by the conscious mind assigns black and white values to our experiences. Therefore, the reality created by the conscious mind depends upon the individual's perception of the 'whiteness' of an object or event. And 'whiteness' can only be determined by comparing it to its 'blackness.' So if any of the concepts to be presented here seem challenging, it may be because you see it as absolutely white, in which case, you need to step back and consider its 'blackness' to get a wider perspective of truth. ***THE** Reality* is found in the grey area, *between* our concepts of black and white, where all possibilities exist.

Chapter I

Follow the Signs

I was always an apprentice of life, passive and objective…a listener. My religious and spiritual beliefs were repeatedly validated as a child and I was comfortable in my relationship with God. But, in my mid twenties, I had an encounter that deepened my appreciation for life and solidified my faith.

I distinctly remember that it was an especially frigid winter evening that I caught my girlfriend cheating on me (again) and I parked my car in a remote corner of a 7-11 convenience store to have a cup of coffee and ponder my situation. As I attempted to achieve an objective assessment of my relationship, I found myself evaluating my entire life to that point. The aspiring musician was getting older and had forfeited several opportunities for success. I was working for very little pay as a diesel mechanic and was hiding my dissatisfaction with myself by spending what little money I earned on drugs and alcohol. I still lived at my parents' home, had no money saved and no meaningful relationships in my life. As I became more self-absorbed, the cherished bond with my family had been steadily deteriorating and it did not appear that I would ever be blessed with a loving relationship and a family of my own. It seemed evident to me that I had peaked in my accomplishments some years earlier and had nowhere to go but down. I had nothing more to offer and it seemed senseless to go on spiraling closer to an undesirable and lonely end. I had served my purpose; therefore it would be prudent to end my life before I would establish relationships that would make my eventual demise more difficult for others to endure. I was strangely comfortable with this evaluation of my life to that point and suicide actually made sense!

As I calmly and deliberately contemplated the means for my departure, a car barreled into the parking lot. At that moment, a woman was leaving the store and a young girl jumped out of the car and ran to her, frantically waving her arms. A minute or so later, the girl got back into her car and left the woman standing alone in front of the store. I was in my black car and in a dark corner of the parking lot and did not realize I was visible, but the woman hesitated only briefly, proceeded across the lot to my car and tapped on the window. I rolled down the window and she said, "Did you see that?"

I giggled... "Yeah, what was her problem?"

She looked confused and said, "She told me that she was at the tabernacle up the block praying and Jesus spoke clearly to her, telling her to go immediately to 7-11 because someone there needs to know that He is with him and that there is more to do. It is not time to leave."

Message received. At that instant, my faith was catapulted to previously inconceivable heights. All at once I realized that whatever happens in my life is not about *me*. My purpose is simply to be here. So I could live my life without fear or doubt; trusting and knowing that I am part of something infinitely greater than my *self*. And my prayers changed from requests for mercy to declarations of gratitude.

About five years later, I had gone to Florida to take care of my grandfather who was ill and alone since my grandmother had passed away.   Living with him in a retirement community made me consider the value of learning life saving techniques and I promptly enrolled in the nearby community college paramedic program.   My grandfather was doing extremely well and shortly after receiving my EMT certification, he insisted that he could take care of himself and that I should make some attempt at getting on with my life. The fact that I was over thirty and had never been married baffled him. He once said, "Things are different today from when I was a kid. Marriage isn't a big thing anymore. You've been going out with a girl for ten years! Marry her, already! If it doesn't work out you could always get rid of her and get another one."   Interesting perspective, but I never married her.

So I packed all of my worldly possessions into a truck and headed to New York with the intention of opening a restaurant of some sort.   Immediately after crossing into Georgia, I ran hard and fast into a severe thunderstorm. Before I even had the chance to slow down, a car from the southbound lanes crossed the grassy divider and was air born right in front of me. I can still see the terrified look on that drivers face as we acknowledged our impending collision. I swerved toward the median and we barely passed each other without touching, but I had just driven my truck into a muddy trough at 65 mph.

The truck rolled several times and settled on its roof. It then continued to spin around and slide another 30 to 40 yards.   Miraculously, (I love that word!) the section of meridian that I had entered is the only section along the interstate devoid of trees. The truck stopped, on its roof, under a foot of mud, about ten feet from the trees. I climbed out into the surging rain and saw more than twenty cars strewn on both sides of the highway. In the half hour or so that it took for the state troopers and paramedics to arrive, I had already stabilized several of the injured. While loading a person into an ambulance, I heard a trooper say to another, "Did anyone check that truck in the divider?" The other trooper replied, "I don't know, but nobody could have survived that."

I had sustained only a small bruise on my shoulder and hip from the seat belt, but I knew I would need a couple of days to straighten out the situation with the totaled rental truck. That night I had called my parents. Not to worry them, I told them I had decided to stop in Atlanta for a day or two to visit with a friend of mine who moved there shortly after high school. The next day, I cleaned out the truck and drove to Atlanta to see my friend and recover from the previous days' trauma. As he and I went through my belongings, we discovered that everything I owned was destroyed. The only thing I could salvage was my school records. As we spread them out on the sidewalk to dry, I remembered that I had heard there was a good chiropractic school in Atlanta. I asked to borrow my friend's car and brought my school records to the Deans' office the next day.

I told the Dean that it was time for me to get my life in gear and I was on my way to New York to open a restaurant unless he would allow me to enroll in the next class which would begin in three weeks. I told him, "I know I can do this. It's up to you. If you let me in, you will not regret it...I'll go all the way...I'll be the best chiropractor I can be." It was obvious to me that I was meant to be there. I didn't ask WHY, it just felt right.

In September of 1989, I began Chiropractic College without a clue. I had never even met a chiropractor before that time. I couldn't possibly list all of the occupations I previously navigated in my search to find a way to make a living, achieve unlimited personal fulfillment and to help other people. But, up to that point, I felt that performing music and catering food were my best options. Chiropractic was the only thing I had not investigated.

Initially, I was a bit frightened. I had to borrow a lot of money to do this, and I didn't know what to expect. Aside from my paramedic training, I had not attended school for eight years prior (how could I possibly recall organic chemistry and biology?), and my hands were still beat up and calloused from my previous years as a diesel mechanic. On top of that, I was a born and bred New Yorker committing to at least four years in a land of farms and slow moving bank lines. Just another test of faith, I guessed.

As it turned out, I wasn't the oldest in my class, I was able to pick up where I left off in biology, and I found that I could survive in a bank line without incident if I got there early, took an occasional deep breath and used simple distraction techniques. I was awed by the quality and abundance of information. Chiropractic philosophy didn't just motivate me, it blew me away! What concepts! THE BRAIN AND CENTRAL NERVOUS SYSTEM CONTROL *EVERYTHING*...FREE THE *ENERGY*... ELIMINATE INTERFERENCE TO *EXPRESSION*.

Network Chiropractic Found Me

Halfway through college, I received a phone call at three o'clock in the morning from someone I hadn't seen for almost fifteen years. I don't even know how she found my number. She said, "I just had the most amazing experience and I'm calling to tell you that you are going to be a 'Network Chiropractor.'" I told her that I would never get involved with *that*. I had heard that it was voodoo and one could get into serious trouble by even mentioning it on campus. "No," I told her. "I plan on being the best doctor I can be. I am going to take x-rays and diagnose peoples' problems…and probably even wear a white lab coat." Then she said, "Look, I'm just delivering a message…have fun!"

Not long after that, I began adjusting in the student clinic. (Chiropractic students have to perform a number of adjustments on each other before being unleashed on the public in the outpatient clinic.) A young woman came to me by way of another student's referral. She had serious spinal problems in the form of bulging discs and direct nerve pressure in her neck and low back. The muscles in her neck and lower back were in a constant state of severe spasm and she experienced intense pain with almost any movement. I could not understand why she had expected that I could help her. I was just a student adjuster. She told me that she was working for a chiropractor and had seen several others. She also tried physical therapy and orthopedists. No one was able to help her and her co-worker told her to call me because I am caring and gentle…so it wouldn't hurt to try.

Well, I was fortunate in that the college I had chosen offered an array of chiropractic techniques. What wasn't available in the curriculum could be learned through technique clubs. I say that I was fortunate because many chiropractic colleges offer limited techniques and some actually require the students to choose only a few techniques to 'specialize' in.

I had seen this young woman several times per week and within a two month period I had tried every chiropractic technique available to me. Nothing I tried yielded any apparent results. Her muscles were in a state of constant spasm from her head to her tailbone. It looked like this poor girl was destined for surgery, which would mean a lifetime of various medications and progressive physical and emotional problems.

I was ready to accept that I was just not equipped or experienced enough to help this woman. Perhaps no one was. Maybe they were lying to us about how chiropractic can help everyone…about how simple it is to remove the interference in the nervous system and 'turn on' the power. Either way, due to my lack of clinical experience and my fear that I might irritate or worsen her condition, I had decided to concede my mission. The ordeal consumed me as I agonized with her for over two months. I'll never forget that feeling of defeat, or what it was like to try to figure out how I would tell her that I had to give up. I had never given up on anything in my life. Being a typical Gemini, I've accumulated many unfinished projects, but I have never given up on any of them!

I was between classes, contemplating the best way to resolve the situation when I spotted a classmate of mine in the hall. I discretely took his arm and pulled him into an empty classroom.

"Listen", I said. And after I explained the situation, I told him that I needed his help. "I know that you've been doing that 'Network' stuff and I've tried everything else. You have to show me how to do it."

He sympathized with the emotional attachment and determination that was devouring me and he rationalized that Network Analysis was a comprehensive system that could not be learned in a single lesson. He briefly described Networks' ten resonant phases (of which, he was familiar with the first five) and showed me how to technically apply a phase one contact. "When you find the indicated spot on the spine to approach, remember that the contact is no more than a light touch. Don't use too much pressure and do not hold the contact for more than two or three respirations. If you happen to find the correct contact point, it will initiate a release of tension from the central nervous system and her muscles will instantly begin to relax."

That evening, I helped the young woman into the prone position on my portable adjusting table and proceeded to perform a simplified phase one evaluation on her. When I had decided where to make the phase one contact, I leaned forward and extended the tip of my thumb. But, I never had the opportunity to actually touch her. As I approached the point of contact (I was still eight to ten inches from touching her body), I saw her spine move in the form of a wave that surged from her low back up to her head. Initially, she started crying. But, the wave moved quickly and seemed to lift her off of the table. By the time it reached the top, she was not just standing, but literally jumping up and down and fervently laughing.

I remember the shock I felt, followed by fear. Did I break her? No, I couldn't have. I didn't even touch her! I was only able to breathe again when she calmed down enough to tell me that she felt "free". There was no more pain when she moved into any position and I was amazed to find that her spine was more flexible and her muscles supple, as most of the spasms were no longer there.

From that point, she was easy to adjust using any chiropractic technique. She had infrequent and short-lived episodes of pain, and then only when she pushed herself. More importantly, she was able to think clearly and was emotionally whole and at ease. It was then that I realized that I needed to explore 'Network Analysis' and holistic healing…and I proceeded underground.

Learning Network Analysis

Over a period of several months, I had attended two basic level Network Spinal Analysis (NSA) seminars and one workshop and frequently met with other students in small groups in between. We would practice on each other and share our observations and experiences. I was amazed at some of the things I had seen. Simply by applying these light contacts, I noticed marked changes in a persons' posture and demeanor. I saw people twitching and even writhing on the tables. Sometimes, I'd see the wave roll up the spine or someone moving into contorted positions. There were frequent outbursts of tears or laughter. When I would ask, "Why are you crying?" The answer was usually," I don't know." When I would ask, "What are you doing?" They would say, "I'm not *doing* anything...it's just happening."

After months of observing these things, I was starting to get annoyed. I had been getting worked on rather frequently and did not really feel anything during my sessions. I did not get emotional or move at all on the table. I thought this was supposed to work for everyone. I was surrounded by people who were experiencing spiritual and physical transformations...and I felt nothing! I finally reached a point where I could rationalize two possible explanations for this. Either I was so spiritually, physically and emotionally well integrated...so perfect that I had no healing to do, or I was the *victim* of a conspiracy. Well, I knew that no person in this three-dimensional reality is beyond growth through transformation. And because 'Network' was relatively new and what I had observed among the small circle of participants was so strange, I rationalized the conspiracy must be real.

How sad.

Poor me.

I guess they just didn't like me enough to let me into their circle: to share the scheme they devised with me. Why would they do that, anyway? For money? Publicity? Maybe just for fun at the expense of people like me! How sick!!!

It did not take long, however, for my feelings of inadequacy to turn to anger. I invested a lot of time and had secured part-time work as a bartender to pay for the seminars. I paid my dues! The entire situation was very uncomfortable for me. It was against my nature to be negative about anything. I have always been modest, but confident. After a while, I decided that a new perspective would be healthy. The time and money I endowed qualified me to attend an advanced seminar. Maybe I was missing something. But, I had decided that if I were to acquiesce to completing the series of seminars and afterwards be subject to the same uncertainties, I would accept that I had been 'had', bless and forgive those that had wasted my time, and redirect my efforts toward therapeutic chiropractic.

Confirmation

There were a few hundred attendees at the advanced seminar. The biggest turnout I had seen at any academic gathering. I got there early and as I entered the room, several people were already getting worked on. There was unrestrained crying and moaning, and people writhing on portable adjusting tables all around me. The atmosphere was thick with emotions and I noticed intermittent sensations of tingling in my hands or on my head and an occasional wrenching in my stomach.

I lined up behind a teaching assistant whom I knew had been practicing Network since its inception. If anyone could help me see what I was missing, it would be her. The session was brief and simple. No structural chiropractic adjusting. Just a few light contacts. When she informed me that I was finished, I told her that the strangest thing happened to me. "Out of nowhere, I got really nauseous while you were working on me."
She said, "Good!"
"No, you don't understand. I felt fine earlier and I feel fine now. And I haven't eaten anything unusual."
"Maybe you needed to throw up! Look, the next time you experience that while your system is being cleared, I want you to take a deep breathe and throw your head back and let it rip!"

Wouldn't that be lovely in a room full of my colleagues? Well, thank God it didn't happen. I had to just let it go. It was a strange sensation, but it had passed and was obviously unimportant.

I participated fully in the three day class. I observed the participants' activities, heard their testimonials, and endeavored to absorb as much information as possible. Although I found it to be intriguing and all the signs were pointing in this direction, I was still not thoroughly convinced that I was on the correct path. When the seminar was finished, everyone lined up at the tables for one last session before leaving.

As I lay prone on the table, I felt overwhelmed with new information and exhausted. Suddenly, I felt my hind end vertically *lifting itself* off of the table as though someone had attached marionette strings to it. All at once, I felt nauseous. But it did not come on gradually. Rather, it felt as though I were thrust into that desperate moment when a person finds themselves two feet from the toilet, unsure if they will reach it in time. I knew I could not hold it back. I was not even able to look up to see if there was a trash can nearby. As I took a deep breathe, I sensed a wave of unimaginable force that rolled up my spine and slapped me in the back of the head. The next thing I remember, I was crying like a baby. I don't know how loud or for how long. But, the very instant I sat up on the table, I had a revelation…

Since I was very young, I had a multitude of respiratory problems. I dealt with asthma and severe allergies in my youth and suffered with chronic bronchitis throughout my teen years. As I got older, I developed chronic pneumonia. I would get ill at least once or twice a year for weeks at a time. But, believe it or not, I did not think it was a big deal. I was used to it! What I realized in the instant that I sat up on that table was that throughout the previous year, while I was training and being consistently worked on, I had not had so much as a sniffle! The work *did* change my life, behind my back…when I wasn't looking!!!

Actually, it was more like *where* I wasn't looking. My first lesson in quantum physics and healing taught me that conscious reality is analogous to watching a clock. Along the way, I was watching and waiting…expecting *some thing*. When am *I* going to spontaneously laugh or cry? Why doesn't *my* body move around while I'm getting worked on? When am I going to notice some dramatic change in my life?

As it turned out, I just needed to stop looking for it. I was completely unaware that my life *was* changing. I had accepted my chronic respiratory problems as just part of who I am. I never expected that to change, so I paid it no attention. But, the things I did give attention to, the clocks I was watching did not seem to budge. And I then realized that the things I directed my attention to, the things that *mattered* to me became fixated in my life simply because my attention was fixed on them.

In universal (quantum) physics, matter and mass are correlates. The things that matter to you will create physical, mental and/or emotional mass in your life.

Chapter Two

Chiropractic

Modern chiropractic was conceived as an energy-based system of bodywork. The goal was simply to free trapped energy in the body and eliminate interference to the flow of neural impulses to ensure complete mind-body communication.

Even at its earliest stages of development, and with some relatively crude techniques, chiropractors were producing positive results with an array of physical and emotional conditions deemed untreatable by medicine. The American Medical Association had already been established as a formidable political entity (in collaboration with drug and supplement manufacturers) and was threatened by a chiropractors' ability to assist a person with only their hands. How could they possibly regulate or sell *that!*

As a consequence, for the first fifty years or so that chiropractic was being developed, chiropractors were often jailed. The AMA said that they were making claims to treat medical conditions with procedures that were not scientifically substantiated. This led to a divergence within the chiropractic profession. There were conformists who felt that validating the science behind spinal adjusting and aligning the profession with the most powerful organization in the world (AMA) would lead to credibility and public acceptance. Others argued that the proof was in the pudding…that chiropractors needed to develop the art of hands-on facilitation through clinical experience.

Those who believed in the art of facilitated healing were simply labeled 'Quacks'. And those who followed the first premise were promptly inducted into the reductionistic medical model. It is, unfortunately, why the bulk of the profession today has been redefined by the AMA as back pain specialists or musculo-skeletal *therapists*. They have lost sight of the original goal which was to develop a hands-on energy based healing system and find themselves to be little more than physical therapists with big student loans.

At this point, there are mountains of research that explain why chiropractic works. We have also established how well it works in the clinical environment. Now we need to stop anal-yzing the ingredients and get more people to taste the pudding!

Chiropractic College was grueling. It is unfortunate that many have fallen prey to anti-chiropractic propaganda. Few realize that the bulk of a medical doctors' education is focused on pharmacology and toxicology. Let's face it, there are new drugs constantly flooding the population and doctors are handing them out in abundance and in various combinations. Hopefully, they learn enough through continuing education (and not so much by trial and error) how not to kill someone with the incorrect dosage or combination of pharmaceuticals.

By contrast, virtually all of a chiropractic education consists of anatomy, physiology and neurology. This essentially means that any chiropractor knows more about how your body works than the worlds' most renowned neurosurgeon! Besides, to me, a 'specialist' is someone who is only good at one thing. Someone who is so focused, they don't even realize that they are in a box. I'm not saying that specialists or medical doctors, in general, are bad.

Everything has its time and purpose. I am merely pointing out that there is no place for them in *healing*. Medicine was designed for crisis intervention…to keep people from being dead. That is its purpose. It never has, nor will it ever help to improve the overall quality of a persons' life, help them to achieve their greatest potential, or inspire healing.

A good chiropractor aspires to understand the biomechanics behind anatomy and the neurological influence on physiology. Furthermore, a holistic practitioner is not particularly concerned with figuring out what a *part* of the body is doing (or not doing), but how every part of the body is effecting and being affected by every other part of the body. Because every aspect of you is connected and interdependent, this brings us to a key Holistic Principle;

You cannot do anything to any part of the body which will not have a simultaneous comprehensive effect on the Whole Body

In reality, to a true facilitator, the body has no *parts.* The physical body is no more than a holographic image. It is a reflection of your state of mind. This will be easier to conceptualize after we have defined brain function.

Il-lusion

As I proceed, please understand that my obligation to you is to simplify apparently complex concepts into practical information you can apply in your life. In order to do this, I will need to expand your focus…open your mind…get you to see things other than the way that you were conditioned to by forces outside of yourself such as your parents, religious influences and political leaders so that you might appreciate the surprisingly uncomplicated reality wherein dwells the spirit. The Truth can be easily distorted beyond comprehension when the head and the heart are in discord. But, when **what we think** is harmonized with **what we feel**, the Truth is clearly seen through intuition…**what we know**.

The conscious mind is polarized. That is to say that we see everything as black and white. The underlying reality is that our spirit, our soul, our true self exists in the grey area between black and white. In a universe of unlimited possibilities, our conscious mind will discern two probabilities (positive or negative, good or bad, etc.) and tend to accept the one that best fits your perception of reality each moment.

Therefore, in order to get you to see things differently or to make a specific point, I may present you with an 'absolute truth' that is contrary to your conditioned perception of reality. For example, although there are times to when nutrition needs to be addressed, when I first meet someone I promptly tell them that I do not believe in vitamins, minerals, herbs or homeopathic remedies. This will usually create a hiccup in their train of thought and their eyes might cross or glaze over for a moment; a sign that a window in their mind was opened. (The more briefly that window stays open, the longer or more tightly it has been closed.)

When I opened my first practice a new patient would frequently tell me, "I tried that holistic stuff. It doesn't work." But, *I* know that the holistic approach *always* works. So I would ask them what kind of holistic care they had tried and they'd tell me that they saw a naturopath or homeopath or herbologist, or that they had undergone vitamin mega-dose, chelation or some form of breathing or light therapy. When they were finished, I would ask, "But what did you do that was holistic?"

As Henry David Thoreau discovered at Walden, the industrial revolution thrust our economy and technology forward, but at the expense of the individual. By and large, the worth of a life is no longer assessed by what values a person creates for humanity, but by how much money it can produce for someone else.

As a consequence of our economically driven society, we are motivated by fads. Not long ago, the word 'Holistic' was being spoken in an increasing number of circles and was, therefore, deemed marketable. What a great way to access new buyers!!! Just tell everyone that you are holistic! Nobody really knows what it means, anyway. Think about this…if I advertise myself in the phone book as a holistic plumber, I can access a whole new clientele. If someone calls me on it, I will just explain that holistic plumbers use *natural enzymes* to keep pipes clean and prevent sewer back-ups. Maybe I will be able to sell a box of 'pipe and drain wellness enzymes' to each customer!

And what about the fads that effect your health as well as your wallet? There are medical fads such as a new prescription drug that everyone seems to need. Let's see if you can connect the dots…

When I was young, I believe that children were receiving their first vaccinations between the ages of 12 – 16 months. That age requirement has been progressively decreased and the suggested number of vaccinations has been increased significantly. They are now vaccinating babies *at birth* while the neurological and physiologic systems have not yet fully developed or are at some critical stage of integration.

I will never forget the reaction I received when I told my parents that I had decided not to vaccinate my children…
"How can you do this to your children? Did we do wrong by you?"
"Of course not, Mom. You did what they told you would be best for me. But, now we know better."

Childhood immunizations have been definitively linked to a plethora of physical and psychological abnormalities. For some, the intense chemical insult to the nervous and immune systems may not create sudden or evident death or disability. The initial reaction may be more subtle. As a child grows and the effects of the chemical trauma they sustained become more obvious, they may manifest post-traumatic symptoms.

But not to worry, we have drugs!!! If a child is having a little trouble concentrating or following orders, we give him Ritalin! If a young woman has painful or irregular menstrual cycles, we can give her hormones! (Etc, etc, etc…)

When you make the rules, you can change them whenever you want. Not too long ago, a total cholesterol count of 350-500 was considered high. Potential profits and effective marketing techniques have, in a very short time, infected the population, creating a cholesterol conscious society. Now, your total cholesterol is considered high if it is over 150. That means that just about everyone you know should be taking cholesterol medication. I guess we should ignore the fact that the laboratory research includes an unacceptable mortality rate associated with these drugs. The truth is cholesterol is not bad at all. It is an inert substance which means that the cholesterol in an egg has no adverse effect on your physiology. So let us consider the cholesterol that your body makes.

We know that the cholesterol molecule itself is inert, so what is it doing floating around in our blood? Could it be there for a reason? The AMA would have us believe that God implanted this substance in your body to make your life miserable, but what if cholesterol was a good thing?

Our arteries are comprised of circular muscles that contract to pump the blood throughout the body. If your blood pressure and heart rate is consistently high (even when we are in a resting state due to tension), the muscles begin to fatigue and the inert cholesterol molecule becomes activated so that plaque can be laid down to support the weakening arterial wall and protect against rupture.

So high blood cholesterol is just another symptom, a sign that your diet is rich in fats and low in fiber, or that your nervous system is anticipating a need to protect the circulatory system. And arterial plaquing is a sign that your circulatory system is under duress. You may not be getting enough exercise/oxygen or your tension levels are high, which can be rectified by meditation and bodywork.

Surgical fads are interesting. They seem to run in cycles. For several years, everyone I met needed tubes put in their ears. Then hip replacements had a hot run, followed by knee replacements. I think that that the replacement hip and knee stockpiles are low because all I hear about now is that there is a waiting list for rotator cuff surgery on the shoulder.

There are also nutritional fads. They come through my office in waves and in many forms such as Noni juice, Kombucha mushrooms, blue-green algae, St. Johns Wart, etc. as well as an array of vitamin, mineral and enzyme supplements and cleanses.

If all of these miracle products actually worked, shouldn't we be a happier and healthier society by now? The TRUTH is that they *do* work. They are designed to make people money and that is exactly what they do. I have scrutinized hundreds of products and, in a few cases, I have actually found some that are properly processed with good (low) concentrations and combinations of ingredients until I notice that they contain Ginko or B-vitamins or Ginseng, etc. I have spoken with many laboratory and company representatives to give them my assessment of their products. When I ask them why they chose to include Ginseng, for example, the answer was always the same…*marketing.* Everyone has heard that Ginseng root is good for you, but most people are unfamiliar with the other ingredients, so it helps to sell the product.

So why does everyone tell me these things are good for me? Could it have anything to do with the several **BILLION** dollars in profits that these companies made last year? I think a more appropriate question to ask is, "Do you think they eat what they are selling?" I don't think so. They see the research reports and are aware of the adverse side effects and long-term damage that saturating your system with toxins can incur. Oh, did I say *toxins?* But herbs and blue-green algae are *natural* so they can't hurt you, right?

I know a wonderful herbologist who was irritated by the self-medicated people who would come to her for help. She would ask them why they were taking Ginseng root and they would say, "The kid behind the counter at the health food store or the cashier at Walmart told me it was good for me!" At one point, she published an article explaining that there are seven different types of Ginseng found in separate geographical areas that have different biochemical effects on the body. If you do not take the correct type in the proper concentration according to your specific chemical constitution, it will adversely affect psychological and physiological function.

There is an abundance of propaganda surrounding the supplement industry such as:

- Vitamin C is water soluble so you can take as much as you want and it won't hurt you. As a matter of fact, the more you take, the healthier you will be. NOT TRUE.

- B-vitamins will give you more energy and should be taken every day. NOT TRUE.

- We need to supplement trace minerals into our diet, because we have depleted our soil resources. NOT TRUE.

- You can prevent osteoporosis by taking calcium. NOT TRUE.

…and I could go on and on and on.

What *is* Holistic?

As its' foundation, the single most important holistic principle states that the body has a natural *tendency to heal itself*. That is what it does, all the time, *if it is allowed to*. It does not matter what or when you would like to heal. As a matter of fact, the things we consciously do to try to force our bodies to heal get in the way of the healing process.

The holistic perspective is quite contrary to that which we have acquired through conditioning. The drug and supplement companies (who control the most influential political power in the world, the American Medical Association) consistently bombard us with messages that infer our dependence on their products and services. How could we survive or lead normal lives without them?! Have we forgotten that the *only* source of healing comes from within? There never was, nor will there ever be, a pill that *heals*. We are so immersed in the illusion created by the money mongers that we do not realize that each time we take that pharmaceutical or vitamin, we are actually conveying an element of our faith from its source within us to a thing outside of us. We need to put our faith, one hundred percent of it, where it belongs and not in the doctor or the pill.

When we go to a therapist or medical doctor for help, we expect them to tell us what is wrong. How negative! Maybe they will give it a strange name or tell us how much worse it could get. Either way, we also expect them to sell us the pill or the shot or surgery to fix it. On a regular basis, I am presented with a new practice member who may have been experiencing symptoms for six months or even six years. Immediately following the consultation and their first session, they will acknowledge that they feel different or that their symptoms have eased for the first time since they could remember.

"Now that we have initiated the process", I would say "I'd like to see you again within three to five days to discuss any changes that you notice in your life and evaluate the physical changes that will have taken place. Can I see you again on Friday?"

"I have an appointment with an orthopedist on Friday, but I might be able to come back on Monday."

"You have already experienced noticeable improvements after only one brief initial session. Considering that you have been in this situation for so long, could you postpone your appointment with the orthopedist and let me work with you for just a week or two? You might be surprised with your progress and find that drugs or surgery are not necessary or, at least, not a worthy pursuit at this time. After all, that should be the very last resort."

"That doctor is so busy that I had to make my appointment three weeks in advance. Besides, I'd like to see what he has to say."

Now, after evaluating this person and observing how they respond to the first session, I often know exactly what the orthopedist would say. I also know what he will offer in the way of treatment. He is going to offer one of his specialized services. You don't go to a hardware store expecting to buy a steak. Orthopedists sell names, pain killers and anti-inflammatories, and surgery. Those are the products and services that they provide. But, is that what you are looking for? Do you need someone to diagnose you and tell you how screwed up you are and how inept you are at healing or living your life without outside assistance before investing adequate time and effort in holistic healing modalities?

I never discourage anyone from seeking diagnosis when it is indicated. But, premature diagnosis or preventative diagnosis leads to an overwhelming number of unnecessary medical interventions. The medical profession is famous for creating and manipulating statistics that support their expensive and growing repertoire of treatments. To keep our focus on the possibility of disease, the general public is frequently reminded of the horrors of breast cancer, but few are aware that there is a direct correlation between the dramatic rise in the appearance of breast cancer and the number of mammographies that are being performed. Basically, if you keep looking you are bound to find something sooner or later. Although it has been established over and over again that the vast majority of situations will resolve with facilitation or that they may resolve spontaneously (the placebo is still proven to be the most effective treatment in clinical trials), there is an imposed urgency to attack what may be a benign or transient condition. It is a shame that I am all too frequently faced with a patient that has come to me as a last resort. They have been repeatedly diagnosed and infused with fear and chemicals which have significantly reinforced and inadvertently accelerated their dis-ease process.

Facilitation

Personally, no matter how hard I might try, I cannot find anything wrong with anyone! All I see is potential…potential for growth through change. In order to assist someone to comprehend the vast difference between the medical and holistic models, my first interview with a new practice member goes something like this;

"I am a hardcore holistic practitioner. I do not believe in vitamins, minerals, herbs or homeopathic remedies and I do not play with gemstones, crystals or magnets. These are all powerful *therapeutic* tools that can interfere with the healing process. From the holistic perspective, all symptoms are simply an expression of tension or interference in the central nervous system (the brain). So whether it is a pimple or a tumor, my job is to get you breathing deeper and to assist your brain in releasing the tension behind the symptom. That is what we call healing. It is always trying to happen. There is nothing you need to do. As a matter of fact, the less you do makes my job easier and your process more comfortable.

I utilize a system of facilitation I call 'Integrated Bodywork'. Although my primary focus is chiropractic, I also incorporate an array of energy- and body-work techniques such as acupressure, polarity, reflexology and much more. Initially, I will apply some light touch contacts. You may or may not feel me working on you, but your breath will deepen, your muscles will relax and your mind may wander. If I do adjust your spine (chiropractically) I may position or gently lean on your body, but I will not hurt you. I do not force anything. If I have to force it, it doesn't belong there.

These are my rules; you must keep your eyes closed and do not talk while I'm working on you. I want you to be comfortable, so do not lay still for *me*. If you feel a spasm, pain or tingling, do not point to it, rub it or tell me about it. I can feel it (literally). Just take a deep breathe and sit up or do a headstand, laugh or cry…whatever feels good. The worst thing that can happen to you is that you will be breathing deeper and you will feel relaxed when we are done.

Your life will immediately begin to change once we have initiated the clearing of tension from your system. The changes you will experience over the next few days or weeks may be subtle or very dramatic. Do not assume that anything physically or emotionally strange or painful that comes up is a new problem. Make note of anything you would consider to be *un-usual* and communicate it with me as it can indicate your dominant stress patterns and tendencies.

If your nervous system releases a lot of tension in a short period of time, we call that a discharge. If you experience a physical discharge, you might feel pain or spasms where you've never had them before. Rapid chemical changes may elicit allergic reactions, shingles, rashes and boils. And with emotional discharging you may notice a sudden shift in your attitude or you could blurt something out and wonder, 'who said that?' Also, flu-like symptoms may arise as a manifestation of resistance."

Do Your Part

We have been trained to present a doctor with our symptoms and leave it up to her to figure out how to fix us. We have willingly relinquished our power to the medical doctors and therapists to the extent that doctors are surprised and even insulted if you were to question their diagnosis or treatment plan. We do not interact with the doctor, we simply trust (put our faith) in him. We literally ask him to find something wrong and accept his diagnosis (or word) as absolute law, and do what we are told out of fear.

The holistic healing model dictates that you must participate in your healing process. The facilitator is nothing more than a catalyst. He is not a healer but rather facilitates (makes easy) your ability to heal.

From a holistic perspective, there are only three things a person needs to do to inspire and support healing. That is to say, there are three critical components to being happy, healthy and living up to your greatest potential:

1

Proper posture and breathing – Do not sit up straight. It looks pretty, but it is detrimental to your health. Your spine is in the shape of a spring for a reason. When you are slouched and in the resting position, the spring is collapsed and your lungs are, for the most part, empty. Inhalation takes energy and as you draw your breath in, your spine straightens, your lungs expand and every cell in your body swells. Exhalation involves no energy or thought at all. It is a matter of letting go, releasing the energy that you drew into the system. When you let go, your veins and lymphatics pump, your lungs empty and your spine collapses as a spring into its flexible resting state. Therefore, when we breathe full, unobstructed breath the spine pumps up and down as the rest of the body expands outward and contracts inward. The combined motion creates respiratory waves that move fluids and nutrients as they ripple throughout the physical body. The movement generated by the respiratory waves is also crucial to maintaining the healing process in that energy in the form of electromagnetic waves is similar to water. When the temperature is below freezing and water flow is restricted, it will freeze. But, if the faucet is opened, even slightly, and the water is allowed to move, it will resist freezing. Likewise, as we breathe more deeply the respiratory wave provides increased motion and support to energy waves so that they may flow more freely. As the breath progressively deepens and resistance to movement decreases, the energy waves gain momentum which accelerates the healing process.

2

Meditation and/or Prayer – Those who do not find the time to meditate or pray *every day* are constantly distracted and preoccupied by *things*. Things that were done, need to be done, should be done or could be done become the primary focus as they *get through* each day. When our attention is focused on what happened yesterday or what might happen tomorrow, we cannot appreciate where we are today. As we invest more time in our personal reality, the world that we create for ourselves, we progressively lose sight of the TRUE reality which binds everything in the universe. As we gradually lose our sense of purpose, our personal world gets smaller and smaller. We become victims of past events and begin to lose hope and motivation with respect to the future because things don't work out the way we *expect* them to. Life becomes confusing and hard work as the tension builds within us. The only opportunity your poor brain gets to relieve your tension and straighten out your twisted perception of reality is when you are asleep and your conscious mind is 'shut off'. You are in a downward spiral as increased tension compromises your breathing which makes it impossible for you to get the deep quality sleep you need to process all that tension.

Meditation gives your brain the ability to process subconscious information while in your awake (conscious) state. If your mind is constantly preoccupied, it shuts off or blocks subconscious information such as intuition which is trying to surface. Remember that, just like healing, meditation is a natural process that is always trying to happen. And it does happen when given the chance. Consistent meditation breaks down the barrier between the conscious and subconscious aspects of the mind to allow Gods' wisdom to shine through you.

3

Bodywork – At one time or another, we have all been in a bad relationship. Everyone around you, your friends and family, are telling you to abort, but you think they just don't want to see you happy. You get defensive and start yelling, "I waited my whole life for this person! You have no right to judge him! I would die without him in my life!"

Then, six months later, you are kicking yourself in the butt and saying "How could I have been so stupid? How come everyone could see it but me?" The answer, of course, is that you had created a reality, a pattern, around that relationship. When you are in a subjective pattern, you cannot be objective. You can only see from the perspective of your subjective reality. As we will explore later, these patterns are established as an inherent function of the nervous system.

So the facilitator can further be defined as an objective party, an outside observer whose purpose is to help you become aware of things in your life that you may not notice because your conscious mind does not consider them part of the reality you are creating. When you understand this truth, you must accept that no one can do life alone. As we become more invested in our life patterns, we become more alone, more individual, more separated from our spirit which connects us to everything. Trapped in our mind, our personal perception of reality eventually considers only *my* life, *my* problems, *my* disease, etc. The message here is one of the most important lessons in life. We depend on each other.

Bodywork, in the form of massage, Reiki, chiropractic, acupressure, etc. is not a luxury. It is critical to breaking our patterns which helps us to maintain a level of objectivity, thereby expanding the possibilities in our lives. Everyone requires consistent bodywork. When a person realizes that their life could be better, that they would benefit from physical, mental and spiritual changes and decides to begin a wellness program, it is important that the frequency of bodywork sessions is substantial at first. In my personal practice I have observed that, for someone who is new to holistic facilitation and not committed to consistent prayer and/or meditation, after we initiate the healing process they will have a tendency to digress to another old comfort zone (habit pattern) within about three to five days. Therefore, I ask new practice members to commit to at least two to three days per week for one month. In that way, I can efficiently monitor their tendencies and keep them processing (changing). As I get to know them and they become more comfortable with me, their process is gaining momentum. The timing is different for everyone, but if the nervous system is frequently cleared the healing process accelerates as the breath and energy movement intensify. As central nervous system interference clears and a person's resistance to change decreases, the process eventually becomes self-sustaining and the frequency of bodywork sessions may be decreased.

It is at this point that a person might be considered to be in a wellness maintenance situation. This means that their nervous system is at a high level of function, they are physically and mentally more relaxed and flexible, and they are more objective and having fun participating in life. Facilitation becomes easier and more effective because the brain is more responsive and it is necessary if this person is traumatized or when accumulated daily tension overwhelms the nervous system, making it less flexible and less aware. My personal and clinical experiences have proved to me that even a person whose nervous system is relatively clear should commit to some form of bodywork at least two to four times per month.

Chapter Three

Control Freak? Me?

Healing is so easy; even a caveman could do it! You don't need to try and it takes no effort. As I previously mentioned, the things we do to figure out or fix our problems actually reinforce them. For example, affirmations could kill you! It doesn't matter how many times you say, "I forgive", or "I have faith", or "I love myself". If it were true, it would not have to be affirmed. You would not need to waste your time and energy trying to convince yourself of something if you believe it to be true. It would be part of your core values. It would be who you are and it would be reflected by your attitude. Moreover, by repeating "I love myself" over and over again you are actually affirming the fact that there are things that you don't like about yourself.

It all boils down to the control issues established in the earliest stages of our development. (If you ever get the opportunity, and are so inclined, you should take a course or read a book on embryology. It is an absolutely fascinating subject.) Our spine is in a C-shape when we are born and the dominant aspect of the brain is parasympathetic. We are primarily under autonomic control. The parasympathetic nervous system inspires us to cry, sweat, poop, etc. We have absolutely no 'sense' of control. Think? We do not have to think. We enter this world as emoters, pure beacons of light and expression, spewing Gods' love and compassion everywhere (for that is our purpose), sensing vibrations in our environment and spontaneously responding to them.

Suppose a mom decided to change her babies' diaper at the wrong time and got peed on. Everyone in the room explodes with laughter at the situation. The baby feels the surge of emotion which unleashes a spray of colors and happy sounds into the world around her. What a wonderful, exhilarating experience!

Less than a year later, the sympathetic (control) aspect of the central nervous system starts to surface. The cervical and lumbar spines form as we learn controlled movements like crawling, walking and talking. Coincidentally, we are laying down the foundation for our control issues as we are taught that we should not be spontaneously expressive. We are told things like, "You have to pee-pee in the potty", "Lower your voice", or "You shouldn't cry unless you have a reason".

The babies' world is changing. I need to *think* about what I'm feeling? I need to be self aware…self conscious? O.K. Maybe that could be interesting and fun. With a sense of self, *I* could create. I like that!

But, I have a problem with the other thing you are telling me. Do you mean I can't just express what I feel? I have to *think* about how I convey my feelings? I need to learn how to hide or ignore what I'm feeling? I can't see how that could benefit anyone, but, I guess, if those are the rules…

And now we have successfully transformed our little emoter into a remoter. Our little beacon of light who was incessantly sending energy impulses out to the universe is now filtering and holding back some of those impulses. One of the rules of facilitation dictates that 'where your attention goes, the energy flows'. And when our attention is directed inward toward the self, our energy field collapses and we become more and more aware of the self and less aware of our ability to feel and heal our environment.

The impression we are left with after the conditioning is that we can control not only the way we express ourselves, but that we can control everything! Meet the Devil…your ego. The *me* …the part of your conscious mind that separates you from everyone and everything. The ego says, "This is my life. I am in control. I don't need anyone."

In reality, we are all expressions of God, inseparable and co-dependent. The only thing that differentiates you from me is our personal life experiences. Those experiences are under our influence, but not our control. For example, if I were to ask you to explain the mechanisms involved in the task of closing your hand, you might assume; I have a thought…I *want* my hand to close. Then my brain sends a signal (a command) to the muscles in my arm and *makes them contract*, thereby closing my hand.

In truth, that is not at all what happens. Consider that you are mostly comprised of water molecules suspended in flowing streams of energy. Shouldn't every part of your body be in constant fluid motion? Wouldn't all of your muscles be continuously contracting and relaxing in spontaneous rhythms in response to the waves of electromagnetic impulses that flow through you? In fact, that would be the case if it weren't for the inhibitory pathways of the brain and spinal cord. It is the breakdown in function or interference to these pathways that may evoke the symptoms of various dis-ease processes such as Huntington's chorea, Parkinson's disease and others where the body displays spontaneous uncontrolled movements.

If these inhibitory pathways are preventing movement, then the act of closing my hand must be interpreted from a different perspective. In reality, I cannot *make* my hand close. But, if it would be beneficial (intuitively) or if I would like my hand to close, the inhibitory pathways will shut off, forfeit control and my hand is *allowed* to close. This is a powerful healing concept. It doesn't matter if you want it to or wish it would…healing always happens when we relinquish our desire to control and allow it to.

Our personal perception of reality is a construct of the conscious mind as it filters the unlimited possibilities that exist in 'True' reality into a few probabilities that we are able to accept as a result of our life experiences. Therefore, the conscious mind can be considered the controlling, inhibiting pathway that limits our possibilities for enhanced life encounters, spiritual growth and healing. When we allow ourselves to be vulnerable and not in control, these things are *allowed* to happen. Life is effortless and bountiful for someone who is familiar and comfortable with these concepts. But for the rest of us, just looking at the word 'vulnerable' can stir up intense fear.

Stress

Here is another word that will make most people cringe. But, maybe it is not a bad word. Perhaps it is just misunderstood. In fact the word 'stress' refers to the most fundamental aspect of life…change. When we are exposed to stress, that is to say when there are changes in our environment, we are inspired to respond or change accordingly. When there are changes in the intensity and direction of the wind, a blade of grass must change its position or bend in response to that particular stress. So grass has to deal with stress, too! Even rocks have stress because they vibrate (resonate) and respond to environmental vibrations. When a rock is exposed to the sun, its atoms are stimulated and its resonant frequency increases. Our perception of the rocks stress response is that it is warming up or giving off heat.

If the general vibration of our environment increases, our brain might perceive that change in frequency as a rise in temperature. The endocrine system is then initiated and we will sweat as one mechanism of cooling the body or lowering the general vibration of our internal environment in response to the rise in the external environment.

So stress is actually a good thing. It is the way we interact with our environment and with each other. It tells us how to adapt and change with everyone and everything that is going on all around us. I love stress! I want as much stress in my life as I can handle. And the more flexible (adaptable) my nervous system is, the more stress I can comfortably process. Stress, or change, is the dynamic element of life that provides opportunities for physical development and spiritual growth.

So if stress is so great, what *is* the problem? How could stress be so good and simultaneously be the root cause of all dis-ease processes? To answer to these questions, we need a better understanding of how the nervous system works.

Holograms

The brain is not as physiologically well-defined as a kidney or liver. It is described as a piezoelectric gel...a liquid mass capable of transmitting electrical impulses. Let us first establish the fact that your brain has no more intelligence than your colon. Intelligence is expressing through every part of you. It is not housed within your cells, but rather constitutes your essence. You are intelligence! Your brain is an organ. Like every other organ, it has a function. The brains' function, as an organ, is to translate vibrations into a three dimensional reality. In effect, your brain is a holographic projector and the 'you' that I perceive is a three dimensional holographic image. How you see yourself is manifested in that image and projected into the world.

I remember a story about a relatively young man who went to see holistic practitioner. He was debilitated as arthritis ravaged his body, fusing his joints. He told the facilitator that he had tried everything else and all of the herbs, vitamins and therapies didn't help at all. He had decided to try energy healing work as a last resort. Of course, *we* know that facilitation should be implemented first in order to avoid the need for medical intervention...*medicine is the alternative.*

The doctor took x-rays prior to initiating a healing program so as to be able to document any future changes in his condition. After only a few months of care, the man was enjoying more range of motion and a significant decrease in pain in all areas of his body. A set of follow-up x-rays confirmed that his arthritis had almost completely regressed. But, there was something else that the doctor noticed; on the first set of x-rays there was a distinct fracture line in the guys' humerus (upper arm bone) that was no longer there. The mans' history revealed that he had broken that arm when he was nine years old. It is tempting to assume that the physical trauma he sustained as a youth eventually led to, or *caused,* his arthritis. But, a more relevant assumption would be that it was one component of his pathological *process* which involves an accumulation of traumas throughout his life experience.

We like to think of ourselves as solid and our bones as the most 'solidest' part of us. The truth is, although bone is the densest, it is also the most fluid tissue in the body. Consider that your bones are reservoirs and there is a dynamic exchange, a constant flow of minerals and other substances between the bone and all parts of the body. So, if the bone is so dynamic and damaged cells are supposed to heal, what was that fracture line doing in his arm twenty-something years after the event?

What it means is that *he* hadn't healed. He did not process that trauma and consequently carried it with him since he was nine years old. And every subsequent experience in his life was influenced by it. In other words, every time he met someone after that event he would shake a persons hand and present himself (his perception of himself) as Jack…the guy with the broken arm. Before long he becomes Jack…the guy with the broken arm, twisted ankle, fear of heights, etc. Because his brain had not processed that trauma, it was reflected in the holographic image that projects *'this is who I am'* to the world. Through facilitation, the event was processed. His perception of self became clearer and his body changed (healed), no longer reflecting a damaged bone.

If I were to take a holographic image of myself, remove a miniscule piece of my fingernail and blow that up to size of the original image, I would see another complete holographic image of myself. Every part of a hologram contains all of the information of the entire hologram. The apparent fracture is a sign or symptom. The information of the event is not only found in the damaged cells, it is an aspect of the entire holographic image.

Brain Function

If the brain worked in a linear fashion (like a computer) we would need a separate file for every sensation such as taste and smell, etc. that we experience in our lives. That would be an impossible amount of information to store and it would be equally difficult to access with any efficiency.

As a hologram, the brain stores information by association. In this way it can store and access enormous (potentially infinite) amounts of information. When something happens, your brain says, "Oh that reminds me of...", and links that sensation or event to something familiar...one that has been experienced before. This helps us to expand on our perceptions of past sensations. The referral of current stimuli to past experience establishes an energy (awareness) loop. Therefore, the creation of energy patterns or habit patterns is fundamental to nervous system function. If we get fixated or stuck in one of these patterns, it will eventually manifest as a dis-ease pattern.

If the job of the nervous system is to process stress (change), then there is the potential for the brain to be healthier and more evolved if our life experiences are varied. For muscles and bones to be strong they must be subject to constant physical stress or exercise. When we do not use our muscles they break down and lose function. In order to maintain a flexible nervous system, we need to be exposed to dynamic fluctuations of energy frequency and intensity. When things stop changing in our lives our brain loses mass and the ability to function or respond to stress properly.

Alzheimer's and dementia are prime examples of the result of living a rigid or well organized lifestyle with minimal variables. An energy pattern will influence brain function and have an effect on the way a person sees themselves and their environment. You might say that it creates an attitude. Since the body is an expression of how you feel, it reacts by constructing a three-dimensional posture (postural pattern) that will reflect that attitude. An imbalanced posture indicates a distortion of energy distribution throughout the body which involves specific muscles and organ functions. When someone invests an increasing amount of time and energy into a pattern, the postural and functional effects become more obvious.

If a person is frequently drawn into an energy pattern, they will eventually become comfortable there; not necessarily because it is a nice place to be, but simply because they visit so often. As a result, they will conduct their lives in ways that support this pattern or comfort zone. They will sit or stand a certain way, develop organized routines around sleeping, dressing, diet and eating, etc. There are fewer variables…less change…less stimulation to the nervous system. Since the pattern has taken over and is running the show, the brain no longer gets any exercise and begins to break down and lose flexibility and function.

From a holistic perspective, hot dogs are good for you. But, if you were to eat them every day they would become a chronic chemical stress to the brain. Your brain needs variety in order to function properly. The same principle applies to vitamins and minerals, herbs and other supplements, and specialized diets. Anything that you do or eat every day provides the potential to establish a new neurological pattern or feed an existing one.

Genetics

Embryology is one of the most fascinating subjects to explore. All cells of the body go through developmental stages. They first appear as non-differentiated cells; cells capable of sustaining their own life, but with no specific function. Consider that the very first cell created at conception is a brain cell. Its identity is assigned by the DNA as it is the purpose of a brain cell to translate the vibrational codes of the DNA into physical form.

The nervous system is the first to develop as the brain cell multiplies into the control center for the future biological system. The nerve cells are assimilating information from not only its DNA, but from the immediate environment as well. This is why a mother's genetic background appears to have a dominant effect on the embryo developing inside of her.

If a mother inherited a genetic predisposition to diabetes, the tendency to express the dis-ease is information resonating within her. In an attempt to create a perfectly functioning person, the nervous system compensates in some way to suppress or control this tendency. When the nervous system is compromised, the dis-ease may express itself.

The embryonic brain is gathering information in order to create a self-sustaining organism. Included in the information it must process is what types of organs are needed and where they should be located and how their respective cells need to be modified to function accordingly. Other factors which affect the assimilation of that information are vibrations in the environment that must also be processed such as the physical, chemical and emotional challenges that mom may be experiencing along with the tendencies for malfunction that resonate within her. In this way, these predispositions for dysfunction are passed on.

From this perspective, genetic inheritance for dis-ease is acquired. It is not in the original DNA blueprint, but rather it is penciled in and as such, it can be erased. That's right! The genetic tendencies for dis-ease can be nullified via vibrational healing, or the changes initiated by energy facilitation. This is known as *transformation*. It is intrinsic to the healing process and evolution. This means that any dis-ease process need not be expressed and is, in fact, reversible. More importantly, it will not be passed on to future generations.

Trauma

The subconscious mind is in constant communication with universal intelligence and is exposed to all electromagnetic energy information including the increasing amount of microwave radiation that we are flooding our environment with. That's right, even the cell phone conversation someone is having on the other side of the world is being processed by your brain! In reality, the conversation (the expression of ideas) is being processed by the brain because thoughts are electromagnetic impulses. But our technology is now amplifying the intensity of those thoughts, good or bad. It is the conscious mind that filters out all of the noise to create your limited three-dimensional physical reality. The demand on the central nervous system must be phenomenal as we increasingly saturate our environment with this microwave energy. Could this have something to do with the progression of physical and social problems we are enduring?

As I mentioned, the holographic brain can process enormous amounts of information. But there is only so much information that can be processed at any one time. This is called 'limitations of matter'. In other words, there is a limit as to how fast you can physically, emotionally and chemically adapt; a limit as to how much information can be processed before the nervous system becomes overwhelmed. Trauma occurs when the nervous system approaches some critical level of overload.

You must realize that your brain will do anything to protect itself. It will literally sacrifice your physical body to protect its integrity. This is most obvious when considering the autoimmune dis-ease processes. For example, with multiple sclerosis the nervous system may be so emotionally or chemically overwhelmed that it needs to find a way to alleviate some of the tension so that it may assume normal function. It will take a chunk of that tension (unprocessed information) and associate it with a part of the body as far away from itself as possible (which is why these situations usually begin in the extremities) and shut off all sensory information coming from that part of the body. The result is decreased sensation and an eventual loss of function in the involved areas. This is the essence of a conscious block. The brain says I'll put it in that part of my body and ignore it. It disassociates itself with that part of the body as though it is no longer part of you. The nervous system is still there, but the mind is no longer aware of it.

Every dis-ease process is nothing more than a protective mechanism of the brain and can be understood from this perspective. Your body is not attacking you. It does not want you to suffer. And *adding* chemical tension to your system in the form of pharmaceutical drugs, herbs or vitamins can **never** help the situation.

Chapter Four

Vibration

A vibration is a little packet of information. Let's look at a radio transmitter and a receiver. To send out a radio signal, the transmitter will generate an electromagnetic impulse (a vibration) which will ripple out in all directions. The properties of the ripple caused by the traveling impulse are determined by the frequency of the vibration and the ripples look like waves. In fact, this is a radio wave...a low voltage electromagnetic energy wave. If the receiver is tuned into the channel or frequency of the wave, we hear the information (music or voice) that the impulse is carrying with it.

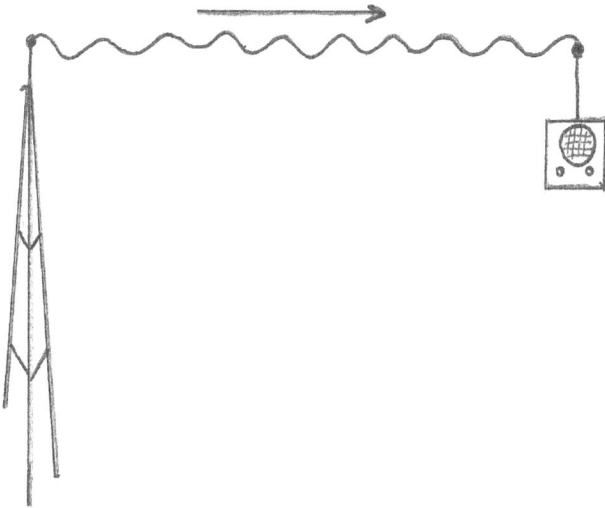

When two traveling impulses are compatible or in phase with each other, they will amplify and intensify each others' signal.

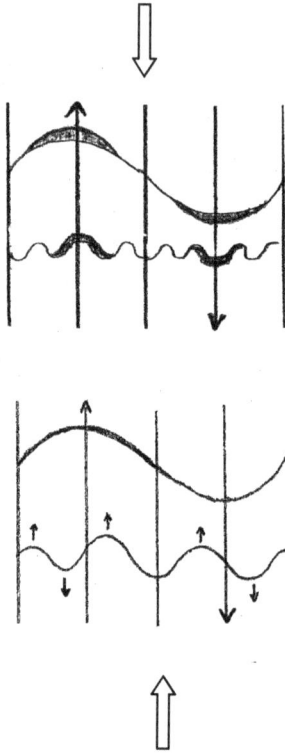

When two traveling impulses are not compatible, they will distort or interfere with each others' signal.

When two incompatible waves cross paths, there is literally a collision of two forces that occurs. Just as when two cars collide a car may lose a bumper or crunch a fender which changes the characteristics of the car, the properties of the original impulse are changed after the collision. This is one example of electromagnetic interference. When the frequency of the impulse is altered and we remain tuned into the original channel or frequency, we hear static because we have lost information where the wave was interfered with by another electromagnetic force or charged material.

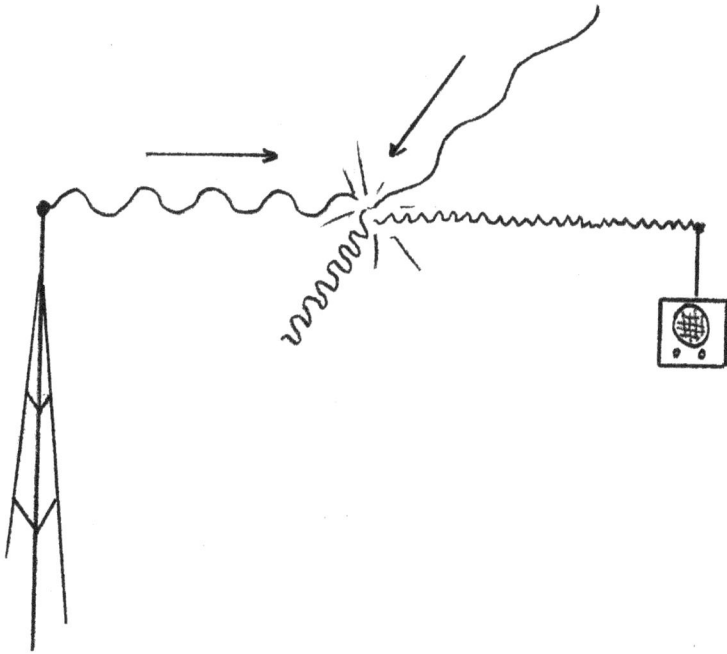

Life Force

An impulse triggers waves that move in all directions simultaneously, but we will give energy direction so that we may simplify and better understand its functional properties. Let's say that these impulses flow into our body through the tail bone, bringing information with them about what is going on in our environment. Our bodies and our brains are bathed with this information and these waves proceed out of the crown or top of the head and are drawn back down, collecting more information as they appear to re-circulate.

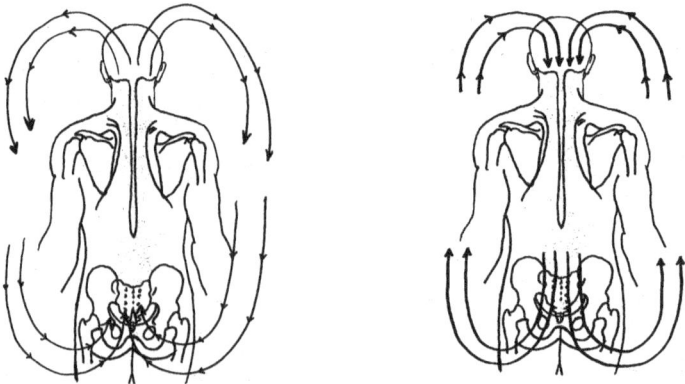

The brain receives this information about what is going on in our environment and its job is to integrate or process it to determine how to best adapt our internal environment to what is happening around us. The brain responds by sending impulses down to all areas of our body so that we will make the appropriate physical, chemical and emotional changes and these waves exit our lower body, expressing our physical response to the universe.

It is this flow of energy waves and the dynamic exchange of information with our environment that generates our energy field. This field is very real and very visible to those whose nervous system is clear enough to perceive subtle energy. In reality, we are all seeing and feeling this energy on a sub-conscious level.

The conscious mind is the aware mind. And awareness can be influenced by desire and expectation...or ego. We focus more on *my* life, *my* needs and *my* problems. As we become more self-conscious (self-aware), our connection with subtle reality (truth) weakens and our attention is increasingly invested in the harsh conscious reality we create. But, the subconscious mind is that of the spirit. It dwells in the light of God...the universal truth. We must take the focus off of our selves, expand our awareness and
break down the barriers between the conscious and subconscious to allow the truth to surface. This is the essence of enlightenment.

As we reconnect more with the spirit, our ability to perceive subtle energy intensifies and the world is once again filled with colors and excitement; just as it was when we were very young.

The Chakras

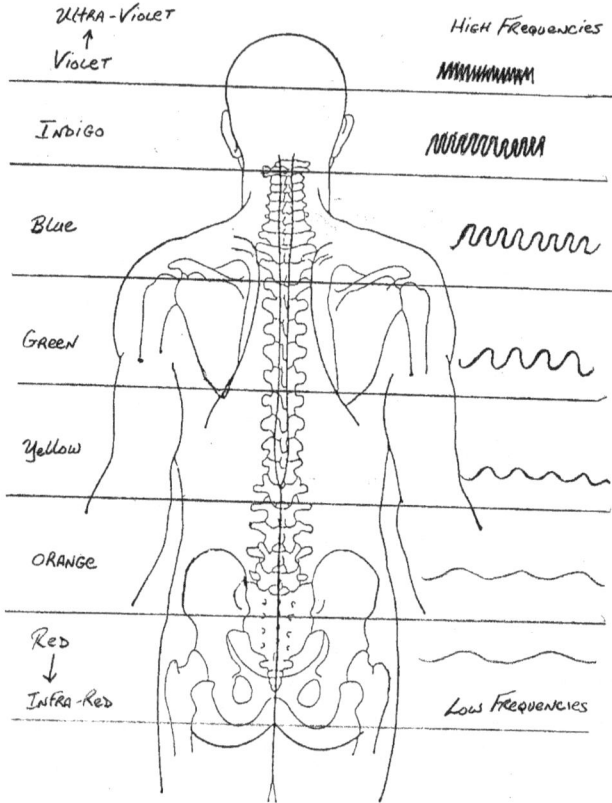

The energy, the life force that animates and flows through us is low voltage electromagnetic energy...radio waves. You are a two-way radio! We are transmitters and receivers of information in the form of electromagnetic impulses or vibrations. If this is true, it would stand to reason that communication is the key to life and evolution.

This means that our one true responsibility is not to ourselves, but to each other. We contribute to creation and the evolution of consciousness when we share our feelings and perceptions using words and touch, music and art, and meditation and prayer. It also means that stifling your emotions and not allowing your thoughts to be expressed is the ultimate sin. When we deny our feelings and our visions, when we separate (disconnect) ourselves from our environment and interrupt the flow of information exchange, it is not only our personal lives that suffer; all that exists is affected.

In many circles, Chakras have been incorrectly referred to as energy centers. You are actually an energy center and the chakras are energy *channels;* the channels that we, as radios, can tune into to transmit and receive specific information.

There are twelve chakras associated with each individual's immediate essence. Seven that lie within the physical body and one associated with each of the five layers of the energy field. Each channel constitutes a specific range of frequencies. If we were to consider the chakras from the tail bone to the head, we can see that the frequencies increase as we move higher. The lowest frequency of light that is perceived by the naked eye is red, which is emitted by the first chakra. And as we progress upward, the colors change with increasing frequencies as they do when we observe a rainbow. The highest frequency we can perceive is violet.

Everything that we consider to be solid reality resonates within this range of frequencies. However, conscious reality is not limited only to what we can see. The conscious minds' ability to integrate information from all of the sensory organs and link to the subconscious or spirit mind gives us the ability to perceive well beyond this range of resonant frequencies.

The vibrant colors of the chakras and the dynamic movement of colors in the energy field (aura) are very real and easily perceived to a person whose nervous system is relatively clear of interference. Remember that color is a vibration that stimulates the optic sensory organs and a vibration is a little packet of information. And color is only one aspect of the information carried by the vibration. In that little packet of information, the color is linked to specific emotions and chemicals that share similar frequencies.

It is also interesting to note that there is an endocrine gland associated with each chakra. Hmmm…doesn't the endocrine system have something to do with hormones? And don't hormones have something to do with emotions? In fact, hormones are called neurotransmitter substances. They are chemicals that are produced by the brain in response to a persons' emotional/chemical state.

Since each of the chakras resonate with frequencies that relate to specific emotions, the hormones that relate to those emotions are found in higher concentrations with respect to the corresponding endocrine gland. The gland does not produce the hormone. The brain produces these chemicals spontaneously when and where they are needed.

For example, it is a medical assumption that a woman's ovaries produce her estrogen. But, estrogen is the chemical that initiates the inflammation response. If I were to cut my finger, estrogen appears at the very instant my skin is compromised. I do not have to wait for estrogen to travel from *my* ovaries to the wound nor is my response to injury compromised because the ovaries were surgically removed.

Therefore, if a gland is surgically removed, hormone replacement therapy is not necessary. In fact, introducing hormones into the body will throw the endocrine system for a loop and adversely affect a person's neurology and perception of reality.

Hormones are the way that your brain translates your emotions into physicality. Another way to say this is that these chemicals allow your body to express how you are really feeling. You may *think* that you are all right with something. Indeed, the conscious mind may block your emotions because they don't fit in with how you want to feel or think you should feel. Although you may not be aware of a conflict, hormones alter your attitude and physiology to reflect your true feelings. It is literally written all over you. This communication happens so that others may recognize and help you resolve your conflict in the interest of conscious evolution. You can interact with someone on a superficial level, sharing the black and white perceptions of the mind. Or you can see (with the heart) between the lines by feeling the chemical and vibrational changes that are inspired by the exchange of information.

To be able to see the truth in others does not require some rare mystical ability. It is a matter of learning how to read the body...and it does require the desire to assist someone along with a willingness to stand in their shoes (compassion).

Energy Blocks

Your perception of reality is defined by how you perceive stress and respond to it as your life unfolds. It can also be affected by the disruption caused by undigested events. This unprocessed information manifests as electromagnetic interference or energy blocks that stunt your physical, emotional and spiritual growth. We are all familiar with the concept that you are what you eat. The materials that you ingest are assimilated by digestion into your molecular structure (body tissues). Likewise, your brain assimilates your life experiences into who you are. This is the essence of spiritual/physical evolution.

Your nervous system can reach a critical point of overload at any time. You don't need to be run over by a truck. You might be traumatized stepping off of a curb or sitting down on a couch. When the brain feels overwhelmed, it reorganizes or resets itself in a brief instant. Although very brief, you might perceive a blackout or a feeling of disorientation or light–headedness. In that moment, the brain takes all of the physical, chemical and emotional components of the event and collapses them into a vibration, which is just a little packet of information. It then takes that vibration and throws it in the back of your mind until there is a more comfortable, appropriate time to process that information. By processing (digesting) the information, our system is given the opportunity to evolve or transform as it would have had the nervous system not been overwhelmed.

But, because there is always an abundance of current information to process; because we are always so preoccupied with what we are *doing* and we may not meditate, exercise or get bodywork on a regular basis, the nervous system may never perceive an appropriate time to bring this undigested information to the surface to be processed. So we carry this vibration around with us until we are able to heal.

Please note that your nervous system may be overwhelmed when good things are happening in your life. *Healing does not pertain to eliminating bad or negative things from your body*...it is simply a matter of fully experiencing your life by processing this undigested information out of the nervous system, releasing you from its effects.

As I mentioned, when the nervous system is traumatized the brain logs the information to that point in your life in a vibration and throws it into the back of your mind. The back of your mind is the tail of your brain, or your spinal cord. That vibration has a frequency and your brain will associate it with an area of the spinal cord (or in a chakra) with which its resonant frequency is *incompatible.* In this way, the vibration creates interference in that channel and in our lives, which makes it more likely to be noticed. If the vibration were associated with an area of compatible frequencies, it would blend in and be less noticeable. If we never notice it, we may never process it and lose the benefits of that life experience. So how does it get our attention?

An energy block is nothing more than this aberrant vibration in the brain which affects the flow of energy through the system. And the uncomfortable symptoms we experience are a consequence of nervous system interference; distortions to the flow of energy which alters physiological function by interrupting communication between the brain and the body.

Chapter Five

Therapy vs. Facilitation

Healing is a process, not an event. When I lecture at massage schools, my primary directive is to outline the principles and the differences between the therapeutic and healing models. I do not discourage anyone from being a therapist...we need therapists. But, if you are going to be a therapist, be the best therapist you can be. Understand that everything has its place, its purpose. Learning *how* to apply therapy is simple. It is a matter of practicing mechanical techniques. Understanding *when* to apply therapy is the most important and challenging concept that can only be learned as the techniques are utilized on individuals in a clinical setting. Unfortunately, because therapy is available in most doctors' offices and is reimbursable by insurance companies, it is grossly over utilized. Instead of trying to find opportunities to use our therapeutic machines and techniques on every person and every symptom, the primary focus should be on when we *need* to use them.

Therapy and facilitation exist in two separate worlds. Facilitation involves the clearing of electromagnetic interference in the central nervous system so as to restore the free flow of energy and re-establish communication between the body and the brain for optimal function.

What we often call therapy has nothing to do with healing. Therapeutic modalities such as electric stimulation (TENS), traction or decompression, heat, etc. do not assist in the release of tension. In fact, they are all ways of adding stress to the body, consequently irritating the nervous system which reinforces and quite often accelerates the pathological process.

It is a travesty that, on a regular basis, a person will walk into my office consumed by a sense of doom. They might look down at their feet, sigh, and in a faint voice tell me that they have 'scoliosis' or 'degenerative disc disease'. They will often get angry when I smile and tell them that it is not as serious as the doctors make it sound.

"But, several doctors told me that I have (own) this 'disease' and nothing can stop it from getting worse and I'll have to learn to live with the pain. How can you say that it's not so bad?! How can you tell me that it can get better?!"

The doctors and therapists treat the symptoms of 'degenerative disc disease' in three ways:

1. Mental/emotional stress – You are trapped in a degenerative process and no one can help you.
2. Chemical stress – Steroid injections, anti-depressants and pain killers.
3. Physical stress – Traction/decompression or surgeries.

Well, it's no wonder why *nobody* gets better with this approach!!!

If symptoms are actually a way that an overloaded nervous system discharges tension, then increasing that tension can never yield healing on any level. The purpose for facilitation is to assist the nervous system in releasing tension, thereby initiating a healing. There is no longer a need to express symptoms and they simply dissipate.

In the world of instant gratification, the magic pills and cures can be misleading.

Let us examine one very common scenario:

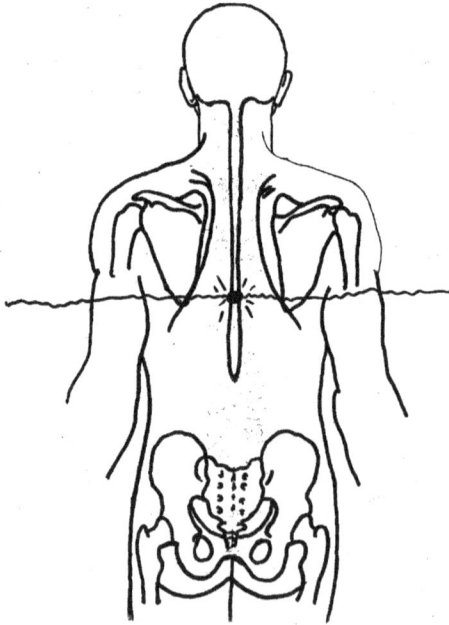

Imagine someone driving home from work on an average day. The illustration above depicts a person whose dominant energy block is located in their third chakra. That block (old information) is trying to come up to the cortex of the brain to be processed. But, the conscious mind is ignoring it because it is too busy dealing with current input like the sensation of the wind blowing on the skin or recent past thoughts of the jerk that almost cut him off or future thoughts (anticipation) of where he is going or what he'll eat for dinner later, etc...

Because the brain is processing so much current information, it may easily receive enough stimulation that it could perceive some critical level of overload. As the driver slows to stop at a red light, he is caught by surprise by the sound of screeching tires and the driver behind him does not stop quite in time and taps the back of his car.

His brain is overwhelmed by the possibilities of what is happening and attempts to log the information to be processed at a later time. During the instant that the nervous system resets, conscious control shuts off and the block surges upward toward the brain. The conscious mind snaps back into action and puts on the brakes. Whoa!!! We're not ready to deal with that NOW!

We all have a tendency for a strong block (resistance) in the throat. After all, it is a center for communication. And we have all learned how to ''swallow' our words when it is inappropriate to express ourselves. So the energy associated with the block surges upward and reinforces the tension in the throat. When the force of the energy at the throat block reaches critical levels, energy is discharged through various channels in much the same way that a dammed river will overflow when its capacity is challenged. This discharge of energy, or electromagnetic radiation, expresses itself and results in nerve irritation and subsequent muscle spasms. *(Which leads to symptoms).*

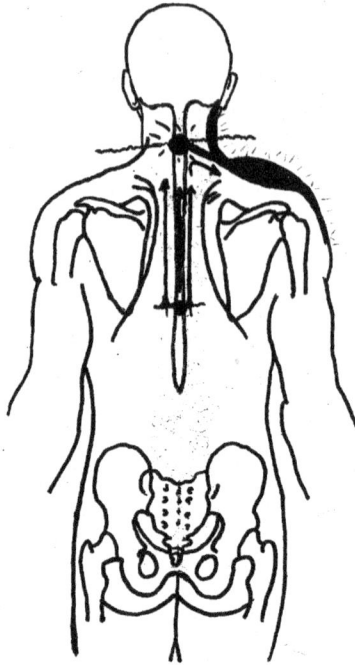

So now this poor person goes to the medical doctor. He might even get lucky and actually *see* the doctor and not his delegated representative. But, more than likely, the doctor will not investigate the patients' body position on impact or the intensity of the accident and will render an 'automatic diagnosis'. Since this patient had a car accident two days ago and he is presenting with a sore throat, rapid heart rate, hot and swollen shoulder muscles and pain and weakness down one arm, he must have whiplash!

Unfortunately, in this case, this person does not have the pinched nerves and torn muscles and ligaments which defines whiplash. Despite that fact, he will be treated for it!

The hot and swollen muscles are serving as pathways by which the nervous system is attempting to relieve tension. A facilitator will assist in the release of that tension at its origin in the central nervous system. But the therapist may attack the symptoms at *their* location by applying heat packs, electric stimulation and deep tissue or trigger point work. This is pure insanity! It is against all reason! Everyone knows that you don't put energy into an overloaded system or it might explode!

By not allowing the nervous system to discharge this tension or by further irritating the situation through the addition of more energy/awareness into those symptomatic areas, we overload the brain with input. Therefore, if this person undergoes a barrage of physical therapy modalities which consistently re-traumatize the brain over treatment periods that often range from six weeks to six months, bad things begin to happen. The person may believe that he is getting better because the pain has lessened. But, in actuality, the brain is trying to compensate for the excessive input by blocking or ignoring it. *Blocking the sensory input is a protective mechanism of the brain at the expense of those areas of the body from which it is disconnecting.*

If you shut off the brains' pressure release valve, the tension continues to build and spill out into other energy channels to find another way to be expressed. The symptoms will move to involve other areas over time. The original irritation was never resolved and is in fact reinforced so that future expressions (symptoms) are intensified.

An example of this is the person who enters my office with uterine fibroids, pelvic arthritis, etc. They have obvious muscle spasms and one shoulder pinned to their ear and a history of intensive physical therapy. When I ask them to tell me about the car accident they were in, they will respond in this manner;

"Oh, that was fifteen years ago! I had whiplash and did physical therapy for about four months, but I haven't had a problem with my neck and shoulder since then. What could that have to do with my low back problem?!"

It is difficult enough for someone to associate their neck with their knee. It becomes virtually impossible to make the connection when there are ten or twenty years between the appearances of the symptoms. Similarly, I often see someone in a state of chemical overload experiencing chemical discharge symptoms or hypersensitivity. Without an understanding of the cumulative affects of tension, the response is *always*, "But, I've been taking synthroid (paxil, zocor, B-vitamins, etc.) for several years and never had a problem with it. Why would I react to it now?"

The holistic practitioner does not focus on any particular symptom, but strives to understand the cause of symptomatic expression by achieving a comprehensive view of the person; considering the physical, chemical and emotional components of their past history and current state of function and to what degree all areas of the body are involved in the primary stress pattern. *Remember, anything affects everything!* By focusing on any one part of the body, we are inadvertently ignoring information that is vital to understanding the source of the dis-ease and its progression over a period of time.

For example, did you know that your head and your hind end are very intimately connected? If someone is suffering from temperomandibular joint pain (TMJ), their first instinct is to go to the dentist. After all, that is the mouth specialist, right? But, a bodyworker who understands musculoskeletal biomechanics, knows that TMJ is most often the result of a pelvic imbalance. If the primary tension is in the pelvis, then the symptomatic TMJ is not the problem. It is a manifestation of the problem far removed from its source. It stands to reason then that if we were to apply surgery, bracing, chemicals such as Botox or steroids, or forced manipulation to the jaw, we would invariably increase the tension in the pelvis and irritate the problem. This would intensify or accelerate the pathological process which would lead to worsened or prolonged TMJ symptoms and eventually force the nervous system to utilize other channels in an attempt to discharge the tension. Five months or five years up the road, the symptoms now manifest as lung tumors, vertigo or arthritis.

It is interesting to observe that in the healing process of the above subject, the patient will often retrace their symptoms as the tension unwinds. I.e. as the arthritis dissipates they may experience episodes of vertigo and TMJ pain that they may not have endured for several years. To the 'specialist' these are distinctly separate problems, each one requiring a different drug or therapy. But to the facilitator, they are components of a single tension which is creating an interference pattern that involves an association between these 'connected' areas of the body.

The 'Now' Moment

From an energetic perspective, we can identify two types of people; the energy suckers and the energy channels. An energy sucker is not a bad person. Keep in mind (no pun intended) that the energy block is always trying to surface from the back of your mind; always pushing up the spinal cord in an attempt to reach the cortex of the brain to be processed. It takes an enormous amount of energy to hold it down. Therefore, if a person has been severely or repeatedly traumatized, or if the initial irritation was not too bad but has been reinforced by physical therapy (mechanical stress) and vitamins, herbs or pharmaceutical drugs (chemical stress) and emotional stresses, much of their life energy is invested in holding this tension in. Energy drawn from all areas of the body is eventually not sufficient and the person's energy field collapses as more is needed to support the conscious block. Since we are all energy generators, another great source of support is other people.

The energy sucker is the type who will irritate you and ask you why you're always pissed off. They do not consciously realize what they are doing. But, if they can get you to vibrate with anger or confusion...to step up your energy output...they will use that extra energy to help suppress their own pent up tension. In essence, they are protecting you and themselves from the unknown. Letting go (vulnerability) can be frightening when you're not sure of what might come up or how you will respond to it (change).

The energy channel, on the other hand, is someone whose nervous system is relatively clear and freely moving energy. They are well integrated and therefore comfortable with themselves and their environment. The aura is dynamic and the energy field is expanded and they have little difficulty expressing how they think and what they feel. They are living in the 'NOW" moment; processing information as it is occurring.

Their perception of what is happening <u>now</u> is not distorted by something they never 'got over' from their past, nor is it altered by some fear of what might happen in the future.

When one of my patients asks me how they will know when they are in the 'now moment', I tell them…"When you look at someone and say 'Wow, you're really pissed off' and walk away, THAT was the 'now moment'. You observed, expressed what you were feeling and then moved on to the next experience.

We know that we are not in the moment when we start pointing fingers…
"Did I make you angry?" "Oh, he must have made you angry!"
These statements indicate blame, guilt and responsibility. Isn't it OK to just resonate with anger or sadness or elation without having to assign a reason for it?

Please understand that conscious choice has very little to do with this. It is a consequence of electromagnetic interference that distorts our perceptions of reality, removing use from the 'now moment'. **In the real world, how you feel is how you feel. It's no ones' fault and it should not be anyone else's problem. It is simply what *is*!**

While making this point in one of my lectures, a woman yelled out frantically, "My God! Could you imagine what this world would be like if everyone said what they felt like saying when they felt like saying it!!!"

The fear in her voice sent me back a few steps. But, my response was, "That is exactly what we are striving for. Wouldn't it be wonderful?"

In reality we are all children of light, expressions of God. Our purpose is simply to communicate. We have been blessed with all of the tools necessary (centered around the endocrine system) to perform this task. We are empathic by nature. When someone's awareness is focused in a particular chakra, we can tune in to that channel to receive their thoughts and feelings. This is always happening on a subconscious level. But, we are often too consciously preoccupied with *my problems* and *what I want* to notice.

As infants, we are stimulated by the vibrations we see and feel around us. As we grow more self-conscious, more concerned with what we are thinking than what we are feeling, we lose touch with the subtle reality. But, *just because we're not noticing it doesn't mean it's not there.* The underlying truth is always right in front of your face waiting to be noticed. And the only tools needed to see it are;

*Compassion and understanding...**receptivity**

*Love and appreciation for life...**gratitude**

*The desire to step out of the box...**objectivity**

*The will to set aside personal desires and

expectation (relinquish control)...**vulnerability**

Conclusions

Anything that your body does it can un-do. That means that arterial plaques and tumors can dissolve, bone deposits can be re-absorbed, damaged or even surgically remove tissue can regenerate, chemical balance can be restored, etc.

Healing is spontaneous and without time and space limitations. Because our pathological dis-ease patterns develop from an accumulation of tension in the central nervous system, it may take many years for them to reach critical levels and surface. But, once the tension is cleared, the nervous system no longer needs to express the symptoms of distress. It may take thirty or forty years for a dis-ease to manifest and all but thirty days or thirty seconds for it to dissipate. It truly is a miracle…the miracle of life. But, most people do not realize that these miracles are inherent in the healing process. In other words, they are normal; that's the way it's supposed to work! Miracles are not reserved for the overtly religious or spiritual or the 'lucky' ones.

Advertising companies for the American Medical Association and pharmaceutical companies have been extremely efficient in brainwashing the general public to believe that we are inept at making our own health choices and unable to survive without their continuous intervention. They have systematically stolen and misplaced the public trust by convincing us that we should put our faith in the doctor or the pill and not in God or our intuition (our selves). Most of us want to trust that the medical doctor knows what is best, but she functions in a complicated world of protocols and malpractice. I think the following line should be added to the Hippocratic oath; '…to protect the financial interest and political stability of the A.M.A…'

I realize that may sound a bit harsh, but why do medical doctors discourage their patients from trying massage, chiropractic, etc? Is it what they believe to be in the best interest of the patient or protocol? Although most doctors would state that reflexology is by and large 'harmless', have they tried it? I have had medical doctors attempt to discredit me (or relieve their frustration) by sending me their 'problem' patients. When these patients respond with a marked improvement in the quality of their lives, the doctor stops referring!

We need to realize that we too often come from a 'lacking' perspective. That is how we've been conditioned because they keep telling us that we need more! This is called 'marketing'. We don't have enough enzymes or minerals; we need more insulin or serotonin; our life would be better with the new Mark III Titanium knee replacement! When the A.M.A. controls medical doctors by setting guidelines for treatment protocol, whose interests are they looking out for? Consider the ramifications if greater than 90% of all medication and surgeries were no longer needed. How would the petroleum industry respond to an engine that runs on sea water?

We are so hard-wired into the Newtonian (reductionistic) model that we have lost our objectivity. We've become too focused on symptoms, isolating one part of the body from the rest. Everyone is a specialist and everyone has a cure. It must be digestive, immune or endocrine…you need more hormones, vitamins or herbs…you need lithium or a hip replacement…etc. But all of these approaches fall short because by focusing on any one system or part of the body, we ignore the adverse effects a treatment may have on other areas or functions.

When applied in a non-therapeutic manner according to holistic principles, facilitation *always* works for everyone, all of the time. There are no exceptions. You do not have to believe it will work for it to be effective. You do not need to try. It is so easy, it's sickening! Simply by committing to a little consistent bodywork, your life will change.

In today's world, most of us spend so much time in front of a television or computer (the marketers' tools) and the rest of our time trying to get more of something. If there is any one thing we need more of, it is contact and communication with other life forms! The truth is, if we feel at any time that we are lacking, what we need is more love and compassion to fill that void.

Increasing our oxygen and water intake and consistent meditation/prayer and bodywork help us to reconnect with our feelings and to express them more easily. The more we get know ourselves (the more comfortable we are with our self), the less likely we are to judge our life process, or other people. We are breathing easier and more relaxed, we are having more fun as we establish ourselves as observers in the life experience, and we are all right with feeling vulnerable...it makes the ride more interesting when, instead of resisting, we yield to whatever moves us.

Life is a generator for change (evolution). As we convert potential energy to kinetic energy, we create new potentials. As I previously mentioned, the nervous system integrates the information of our individual life experiences into who we are. Concurrently, the higher consciousness evolves and the universe unfolds as it assimilates the sum total of all life experience. What I'm saying here is that each individual is a hologram, interpreting and digesting its experiences. And every living organism is a piece of a greater hologram (God, universal intelligence, cosmic consciousness) that evolves by assimilating the sum total of all life experience, reaping the benefits of life itself. Therefore, the trials and pleasures that we encounter are pieces of a large puzzle; fuel for the evolution of consciousness. In effect, most of the things that happen to you are not at all about *you*. And our contribution to creation for the privilege of having the life experience depends upon our willingness to interact objectively with our environment, express ourselves freely, and not take whatever comes personally.

The world, as our parents knew it, is changing. What is not working is becoming more apparent, causing people to start thinking for themselves and to rearrange their priorities. Future generations will look upon the past one hundred years with confusion. They will not be able to conceive of how a civilization that is fragmented and in emotional isolation could exist. That is of course, if we do in fact survive!

Do not let anyone tell you what is right for you...draw your own conclusions. In fact, don't believe a word I've written. Just try it on and see how it fits. What have you got to lose...a tumor...a bad relationship...a little tension..? It most definitely can't hurt and it may even help to make your life more productive and enjoyable.

www.ingramcontent.com/pod-product-compliance
Lightning Source LLC
Chambersburg PA
CBHW031814190326
41518CB00006B/331